S0-AVF-145

PROPERTIES OF ATOMS & MOLECULES

TEACHER SUPPLEMENT

1:1

answersingenesis

Petersburg, Kentucky, USA

ANSWERS IN GENESIS **SCIENCE** BY DEBBIE & RICHARD LAWRENCE

God's Design for Chemistry & Ecology
Properties of Atoms & Molecules Teacher Supplement

© 2008 by Debbie & Richard Lawrence

Third printing: January 2014

Published by Answers in Genesis, 2800 Bullittsburg Church Rd., Petersburg KY 41080

You may contact the authors at (970) 686-5744

ISBN: 1-60092-284-8

Cover design & layout: Diane King
Editors: Lori Jaworski, Gary Vaterlaus

Printed in China.

www.answersingenesis.org   www.godsdesignscience.com

TABLE OF CONTENTS

Welcome to
God's Design® for Chemistry & Ecology

God's Design for Chemistry & Ecology is a series that has been designed for use in teaching chemistry and ecology to elementary and middle school students. It is divided into three books: *Properties of Matter*, *Properties of Atoms and Molecules*, and *Properties of Ecosystems*. Each book has 35 lessons including a final project that ties all of the lessons together.

In addition to the lessons, special features in each book include biographical information on interesting people as well as fun facts to make the subject more fun.

Although this is a complete curriculum, the information included here is just a beginning, so please feel free to add to each lesson as you see fit. A resource guide is included in the appendices to help you find additional information and resources. A list of supplies needed is included at the beginning of each lesson, while a master list of all supplies needed for the entire series can be found in the appendices.

Answer keys for all review questions, worksheets, quizzes, and the final exam are included here. Reproducible student worksheets and tests may be found on the supplementary CD-Rom for easy printing. Please contact Answers in Genesis if you wish to purchase a printed version of all the student materials, or go to www.AnswersBookstore.com.

If you wish to get through all three books of the *Chemistry & Ecology* series in one year, you should plan on covering approximately three lessons per week. The time required for each lesson varies depending on how much additional information you want to include, but you can plan on about 45 minutes per lesson.

If you wish to cover the material in more depth, you may add additional information and take a longer period of time to cover all the material or you could choose to do only one or two of the books in the series as a unit study.

Why Teach Chemistry & Ecology?

Maybe you hate science or you just hate teaching it. Maybe you love science but don't quite know how to teach it to your children. Maybe science just doesn't seem as important as some of those other subjects you need to teach. Maybe you need a little motivation. If any of these descriptions fits you, then please consider the following.

It is not uncommon to question the need to teach your kids hands-on science in elementary school. We could argue that the knowledge gained in science will be needed later in life in order for your children to be more productive and well-rounded adults. We could argue that teaching your children science also teaches them logical and inductive thinking and reasoning skills, which are tools they will need to be more successful. We could argue that science is a necessity in this technological world in which we live. While all of these arguments are true, not one of them is the real reason that we should teach our children science.

The most important reason to teach science in elementary school is to give your children an understanding that God is our Creator, and the Bible can be trusted. Teaching science from a creation perspective is one of the best ways to reinforce your children's faith in God and to help them counter the evolutionary propaganda they face every day.

God is the Master Creator of everything. His handiwork is all around us. Our Great Creator put in place all of the laws of physics, biology, and chemistry. These laws were put here for us to see His wisdom and power. In science, we see the hand of God at work more than in any other subject. Romans 1:20 says, "For since the creation of the world His invisible attributes are clearly seen, being understood by the things that are made, even His eternal power and Godhead, so that they [men] are without excuse." We need to help our children see God as Creator of the world around them so they will be able to recognize God and follow Him.

The study of chemistry helps us understand and appreciate the amazing way everything God created works together. The study of atoms and molecules and how different substances react with each other reveals an amazing design, even at the smallest level of life. Understanding the carbon, nitrogen, and water cycles helps our children see that God has a plan to keep everything working together. Learning about ecosystems reveals God's genius in nature.

It's fun to teach chemistry and ecology! It's interesting too. The elements of chemistry are all around us. Children naturally like to combine things to see what will happen. You just need to direct their curiosity.

Finally, teaching chemistry is easy. You won't have to try to find strange materials for experiments or do dangerous things to learn about chemistry. Chemistry is as close as your kitchen or your own body, and ecosystems are just outside your door.

HOW DO I TEACH SCIENCE?

In order to teach any subject, you need to understand that people learn in different ways. Most people, and children in particular, have a dominant or preferred learning style in which they absorb and retain information more easily.

If a student's dominant style is:

Auditory
He needs not only to hear the information but he needs to hear himself say it. This child needs oral presentation as well as oral drill and repetition.
Visual
She needs things she can see. This child responds well to flashcards, pictures, charts, models, etc.
Kinesthetic
He needs active participation. This child remembers best through games, hands-on activities, experiments, and field trips.

Also, some people are more relational while others are more analytical. The relational student needs to know why this subject is important and how it will affect him personally. The analytical student, however, wants just the facts.

If you are trying to teach more than one student, you will probably have to deal with more than one learning style. Therefore, you need to present your lessons in several different ways so that each student can grasp and retain the information.

GRADES 3–8

Each lesson should be completed by all upper elementary and junior high students. This is the main part of the lesson containing a reading section, a hands-on activity that reinforces the ideas in the reading section (blue box), and a review section that provides review questions and application questions (red box).

GRADES 6–8

For middle school/junior high age students, we provide a "Challenge" section that contains more challenging material as well as additional activities and projects for older students (green box).

We suggest a threefold approach to each lesson:

Introduce the topic

We give a brief description of the facts. Frequently you will want to add more information than the essentials given in this book. In addition to reading this section aloud, you may wish to do one or more of the following:

- Read a related book with your students.
- Write things down to help your visual students.
- Give some history of the subject. We provide some historical sketches to help you, but you may want to add more.
- Ask questions to get your students thinking about the subject.

Make observations and do experiments

- Hands-on projects are suggested for each lesson. This section of each lesson may require help from the teacher.
- Have your students perform the activity by themselves whenever possible.

Review

- The "What did we learn?" section has review questions.
- The "Taking it further" section encourages students to
  - Draw conclusions
  - Make applications of what was learned
  - Add extended information to what was covered in the lesson
- The "FUN FACT" section adds fun or interesting information.

By teaching all three parts of the lesson, you will be presenting the material in a way that all learning styles can both relate to and remember.

Also, this approach relates directly to the scientific method and will help your students think more scientifically. The *scientific method* is just a way to examine a subject logically and learn from it. Briefly, the steps of the scientific method are:

1. Learn about a topic.
2. Ask a question.
3. Make a hypothesis (a good guess).
4. Design an experiment to test your hypothesis.
5. Observe the experiment and collect data.
6. Draw conclusions. (Does the data support your hypothesis?)

Note: It's okay to have a "wrong hypothesis." That's how we learn. Be sure to help your students understand why they sometimes get a different result than expected.

Our lessons will help your students begin to approach problems in a logical, scientific way.

How Do I Teach Creation vs. Evolution?

We are constantly bombarded by evolutionary ideas about the earth in books, movies, museums, and even commercials. These raise many questions: Is a living being just a collection of chemicals? Did life begin as a random combination of chemicals? Can life be recreated in a laboratory? What does the chemical evidence tell us about the earth? The Bible answers these questions, and this book accepts the historical accuracy of the Bible as written. We believe this is the only way we can teach our children to trust that everything God says is true.

There are five common views of the origins of life and the age of the earth:

Historical biblical account	Progressive creation	Gap theory	Theistic evolution	Naturalistic evolution
Each day of creation in Genesis is a normal day of about 24 hours in length, in which God created everything that exists. The earth is only thousands of years old, as determined by the genealogies in the Bible.	The idea that God created various creatures to replace other creatures that died out over millions of years. Each of the days in Genesis represents a long period of time (day-age view) and the earth is billions of years old.	The idea that there was a long, long time between what happened in Genesis 1:1 and what happened in Genesis 1:2. During this time, the "fossil record" was supposed to have formed, and millions of years of earth history supposedly passed.	The idea that God used the process of evolution over millions of years (involving struggle and death) to bring about what we see today.	The view that there is no God and evolution of all life forms happened by purely naturalistic processes over billions of years.

Any theory that tries to combine the evolutionary time frame with creation presupposes that death entered the world before Adam sinned, which contradicts what God has said in His Word. The view that the earth (and its "fossil record") is hundreds of millions of years old damages the gospel message. God's completed creation was "very good" at the end of the sixth day (Genesis 1:31). Death entered this perfect paradise *after* Adam disobeyed God's command. It was the punishment for Adam's sin (Genesis 2:16–17; 3:19; Romans 5:12–19). Thorns appeared when God cursed the ground because of Adam's sin (Genesis 3:18).

The first animal death occurred when God killed at least one animal, shedding its blood, to make clothes for Adam and Eve (Genesis 3:21). If the earth's "fossil record" (filled with death, disease, and thorns) formed over millions of years before Adam appeared (and before he sinned), then death no longer would be the penalty for sin. Death, the "last enemy" (1 Corinthians 15:26), diseases (such as cancer), and thorns would instead be part of the original creation that God labeled "very good." No, it is clear that the "fossil record" formed sometime *after* Adam sinned—not many millions of years before. Most fossils were formed as a result of the worldwide Genesis Flood.

When viewed from a biblical perspective, the scientific evidence clearly supports a recent creation by God, and not naturalistic evolution and millions of years. The volume of evidence supporting the biblical creation account is substantial and cannot be adequately covered in this book. If you would like more information on this topic, please see the resource guide in the appendices. To help get you started, just a few examples of evidence supporting biblical creation are given below:

Evolutionary Myth: Life evolved from non-life when chemicals randomly combined together to produce amino acids and then proteins that produced living cells.

The Truth: The chemical requirements for DNA and proteins to line up just right to create life could not have happened through purely natural processes. The process of converting DNA information into proteins requires at least 75 different protein molecules. But each and every one of these 75 proteins must be synthesized in the first place by the process in which they themselves are involved. How could the process begin without the presence of all the necessary proteins? Could all 75 proteins have arisen by chance in just the right place at just the right time? Dr. Gary Parker says this is like the chicken and the egg problem. The obvious conclusion is that both the DNA and proteins must have been functional from the beginning, otherwise life could not exist. The best explanation for the existence of these proteins and DNA is that God created them.

Gary Parker, *Creation: Facts of Life* (Master Books, 2006), pp. 20–43.

Evolutionary Myth: Stanley Miller created life in a test tube, thus demonstrating that the early earth had the conditions necessary for life to begin.

The Truth: Although Miller was able to create amino acids from raw chemicals in his famous experiment, he did not create anything close to life or even the ingredients of life. There are four main problems with Miller's experiment. First, he left out oxygen because he knew that oxygen corrodes and destroys amino acids very quickly. However, rocks found in every layer of the earth indicate that oxygen has always been a part of the earth's atmosphere. Second, Miller included ammonia gas and methane gas. Ammonia gas would not have been present in any large quantities because it would have been dissolved in the oceans. And there is no indication in any of the rock layers that methane has ever been a part of the earth's atmosphere. Third, Miller used a spark of electricity to cause the amino acids to form, simulating lightning. However, this spark more quickly destroyed the amino acids than built them up, so to keep the amino acids from being destroyed, Miller used specially designed equipment to siphon off the amino acids before they could be destroyed. This is not what would have happened in nature. And finally, although Miller did produce amino acids, they were not the kinds of amino acids that are needed for life as we know it. Most of the acids were ones that actually break down proteins, not build them up.

Mike Riddle, "Can Natural Processes Explain the Origin of Life," in *The New Answers Book 2*, Ken Ham, ed. (Master Books, 2008). See also www.answersingenesis.org/go/origin.

Evolutionary Myth: Living creatures are just a collection of chemicals.

The Truth: It is true that cells are made of specific chemicals. However, a dead animal is made of the same chemicals as it was when it was living, but it cannot become alive again. What makes the chemicals into a living creature is the result of the organization of the substances, not just the substances themselves. Dr. Parker again uses an example. An airplane is made up of millions of non-flying parts; however, it can fly because of the design and organization of those parts. Similarly, plants and animals are alive because God created the chemicals in a specific way for them to be able to live. A collection of all the right parts is not life.

Evolutionary Myth: Chemical evidence points to an earth that is billions of years old.

The Truth: Much of the chemical evidence actually points to a young earth. For example, radioactive decay in the earth's crust produces helium atoms that rise to the surface and enter the atmosphere. Assuming that the rate of helium production has always been constant (an evolutionary assumption), the maximum age for the atmosphere could only be 2 million years.[1] This is much younger than the 4+ billion years claimed by evolutionists. And there are many ideas that could explain the presence of helium that would indicate a much younger age than 2 million years. Similarly, salt accumulates in the ocean over time. Evolutionists claim that life evolved in a salty ocean 3–4 billion years ago. If this were true and the salt has continued to accumulate over billions of years, the ocean would be too salty for anything to live in by now. Using the most conservative possible values (those that would give the oldest possible age for the oceans), scientists have calculated that the ocean must be less than 62 million years. That number is based on the assumption that nothing has affected the rate at which the salt is accumulating. However, the Genesis Flood would have drastically altered the amount of salt in the ocean, dissolving much sodium from land rocks.[2] Thus, the chemical evidence does not support an earth that is billions of years old.

[1] Don DeYoung, *Thousands…not billions* (Master Books, 2005).
[2] John D. Morris, *The Young Earth* (Master Books, 2007), pp. 83–87. See also www.answersingenesis.org/go/salty.

Despite the claims of many scientists, if you examine the evidence objectively, it is obvious that evolution and millions of years have not been proven. You can be confident that if you teach that what the Bible says is true, you won't go wrong. Instill in your student a confidence in the truth of the Bible in all areas. If scientific thought seems to contradict the Bible, realize that scientists often make mistakes, but God does not lie. At one time scientists believed that the earth was the center of the universe, that living things could spring from non-living things, and that blood-letting was good for the body. All of these were believed to be scientific facts but have since been disproved, but the Word of God remains true. If we use modern "science" to interpret the Bible, what will happen to our faith in God's Word when scientists change their theories yet again?

Integrating the Seven C's

Throughout the *God's Design® for Science* series you will see icons that represent the Seven C's of History. The Seven C's is a framework in which all of history, and the future to come, can be placed. As we go through our daily routines we may not understand how the details of life connect with the truth that we find in the Bible. This is also the case for students. When discussing the importance of the Bible you may find yourself telling students that the Bible is relevant in everyday activities. But how do we help the younger generation see that? The Seven C's are intended to help.

The Seven C's can be used to develop a biblical worldview in students, young or old. Much more than entertaining stories and religious teachings, the Bible has real connections to our everyday life. It may be hard, at first, to see how many connections there are, but with practice, the daily relevance of God's Word will come alive. Let's look at the Seven C's of History and how each can be connected to what the students are learning.

Creation

God perfectly created the heavens, the earth, and all that is in them in six normal-length days around 6,000 years ago.

This teaching is foundational to a biblical worldview and can be put into the context of any subject. In science, the amazing design that we see in nature—whether in the veins of a leaf or the complexity of your hand—is all the handiwork of God. Virtually all of the lessons in *God's Design for Science* can be related to God's creation of the heavens and earth.

Other contexts include:

Natural laws—any discussion of a law of nature naturally leads to God's creative power.

DNA and information—the information in every living thing was created by God's supreme intelligence.

Mathematics—the laws of mathematics reflect the order of the Creator.

Biological diversity—the distinct kinds of animals that we see were created during the Creation Week, not as products of evolution.

Art—the creativity of man is demonstrated through various art forms.

History—all time scales can be compared to the biblical time scale extending back about 6,000 years.

Ecology—God has called mankind to act as stewards over His creation.

Corruption

After God completed His perfect creation, Adam disobeyed God by eating the forbidden fruit. As a result, sin and death entered the world, and the world has been in decay since that time. This point is evident throughout the world that we live in. The struggle for survival in animals, the death of loved ones, and the violence all around us are all examples of the corrupting influence of sin.

Other contexts include:

Genetics—the mutations that lead to diseases, cancer, and variation within populations are the result of corruption.

Biological relationships—predators and parasites result from corruption.

History—wars and struggles between mankind, exemplified in the account of Cain and Abel, are a result of sin.

Catastrophe

God was grieved by the wickedness of mankind and judged this wickedness with a global Flood. The Flood covered the entire surface of the earth and killed all air-breathing creatures that were not aboard the Ark. The eight people and the animals aboard the Ark replenished the earth after God delivered them from the catastrophe.

The catastrophe described in the Bible would naturally leave behind much evidence. The studies of geology and of the biological diversity of animals on the planet are two of the most obvious applications of this event. Much of scientific understanding is based on how a scientist views the events of the Genesis Flood.

Other contexts include:

Biological diversity—all of the birds, mammals, and other air-breathing animals have populated the earth from the original kinds which left the Ark.

Geology—the layers of sedimentary rock seen in roadcuts, canyons, and other geologic features are testaments to the global Flood.

Geography—features like mountains, valleys, and plains were formed as the floodwaters receded.

Physics—rainbows are a perennial sign of God's faithfulness and His pledge to never flood the entire earth again.

Fossils—Most fossils are a result of the Flood rapidly burying plants and animals.

Plate tectonics—the rapid movement of the earth's plates likely accompanied the Flood.

Global warming/Ice Age—both of these items are likely a result of the activity of the Flood. The warming we are experiencing today has been present since the peak of the Ice Age (with variations over time).

CONFUSION

God commanded Noah and his descendants to spread across the earth. The refusal to obey this command and the building of the tower at Babel caused God to judge this sin. The common language of the people was confused and they spread across the globe as groups with a common language. All people are truly of "one blood" as descendants of Noah and, originally, Adam.

The confusion of the languages led people to scatter across the globe. As people settled in new areas, the traits they carried with them became concentrated in those populations. Traits like dark skin were beneficial in the tropics while other traits benefited populations in northern climates,

and distinct people groups, not races, developed. Other contexts include:

Genetics—the study of human DNA has shown that there is little difference in the genetic makeup of the so-called "races."

Languages—there are about seventy language groups from which all modern languages have developed.

Archaeology—the presence of common building structures, like pyramids, around the world confirms the biblical account.

Literature—recorded and oral records tell of similar events relating to the Flood and the dispersion at Babel.

CHRIST

God did not leave mankind without a way to be redeemed from its sinful state. The Law was given to Moses to show how far away man is from God's standard of perfection. Rather than the sacrifices, which only covered sins, people needed a Savior to take away their sin. This was accomplished when Jesus Christ came to earth to live a perfect life and, by that obedience, was able to be the sacrifice to satisfy God's wrath for all who believe.

The deity of Christ and the amazing plan that was set forth before the foundation of the earth is the core of Christian doctrine. The earthly life of Jesus was the fulfillment of many prophecies and confirms the truthfulness of the Bible. His miracles and presence in human form demonstrate that God is both intimately concerned with His creation and able to control it in an absolute way.

Other contexts include:

Psychology—popular secular psychology teaches of the inherent goodness of man, but Christ has lived the only perfect life. Mankind needs a Savior to redeem it from its unrighteousness.

Biology—Christ's virgin birth demonstrates God's sovereignty over nature.

Physics—turning the water into wine and the feeding of the five thousand demonstrate Christ's deity and His sovereignty over nature.

History—time is marked (in the western world)

based on the birth of Christ despite current efforts to change the meaning.

Art—much art is based on the life of Christ and many of the masters are known for these depictions, whether on canvas or in music.

CROSS

Because God is perfectly just and holy, He must punish sin. The sinless life of Jesus Christ was offered as a substitutionary sacrifice for all of those who will repent and put their faith in the Savior. After His death on the Cross, He defeated death by rising on the third day and is now seated at the right hand of God.

The events surrounding the crucifixion and resurrection have a most significant place in the life of Christians. Though there is no way to scientifically prove the resurrection, there is likewise no way to prove the stories of evolutionary history. These are matters of faith founded in the truth of God's Word and His character. The eyewitness testimony of over 500 people and the written Word of God provide the basis for our belief.

Other contexts include:

Biology—the biological details of the crucifixion can be studied alongside the anatomy of the human body.

History—the use of crucifixion as a method of punishment was short-lived in historical terms and not known at the time it was prophesied.

Art—the crucifixion and resurrection have inspired many wonderful works of art.

CONSUMMATION

God, in His great mercy, has promised that He will restore the earth to its original state—a world without death, suffering, war, and disease. The corruption introduced by Adam's sin will be removed. Those who have repented and put their trust in the completed work of Christ on the Cross will experience life in this new heaven and earth. We will be able to enjoy and worship God forever in a perfect place.

This future event is a little more difficult to connect with academic subjects. However, the hope of a life in God's presence and in the absence of sin can be inserted in discussions of human conflict, disease, suffering, and sin in general.

Other contexts include:

History—in discussions of war or human conflict the coming age offers hope.

Biology—the violent struggle for life seen in the predator-prey relationships will no longer taint the earth.

Medicine—while we struggle to find cures for diseases and alleviate the suffering of those enduring the effects of the Curse, we ultimately place our hope in the healing that will come in the eternal state.

The preceding examples are given to provide ideas for integrating the Seven C's of History into a broad range of curriculum activities. We would recommend that you give your students, and yourself, a better understanding of the Seven C's framework by using AiG's *Answers for Kids* curriculum. The first seven lessons of this curriculum cover the Seven C's and will establish a solid understanding of the true history, and future, of the universe. Full lesson plans, activities, and student resources are provided in the curriculum set.

We also offer bookmarks displaying the Seven C's and a wall chart. These can be used as visual cues for the students to help them recall the information and integrate new learning into its proper place in a biblical worldview.

Even if you use other curricula, you can still incorporate the Seven C's teaching into those. Using this approach will help students make firm connections between biblical events and every aspect of the world around them, and they will begin to develop a truly biblical worldview and not just add pieces of the Bible to what they learn in "the real world."

UNIT 1
ATOMS & MOLECULES

INTRODUCTION TO CHEMISTRY
THE STUDY OF MATTER AND MOLECULES

SUPPLY LIST

Drinking glass   Baking soda   Vinegar
Supplies for Challenge: 2-liter bottle of diet soda   Mentos® candies   Toothpick   Tape
Piece of paper

WHAT DID WE LEARN?

- What is matter? **Anything that has mass and takes up space.**
- Does air have mass? **Yes. It may seem like there is nothing there, but even though air is very light, it still has mass. The air contains molecules that take up space.**
- What do chemists study? **The way matter reacts with other matter and the environment.**

TAKING IT FURTHER

- Would you expect to see the same reaction each time you combine baking soda and vinegar? **Yes, because God designed certain laws for matter to follow, so we would expect it to react the same way each time.**

ATOMS
BASIC BUILDING BLOCKS

SUPPLY LIST

Copy of "Atomic Models" worksheet   Colored pencils
Supplies for Challenge: Copy of "Energy Levels" worksheet

ATOMIC MODELS WORKSHEET

- Color the protons in each atom red (white), the neutrons blue (black), and the electrons gray.

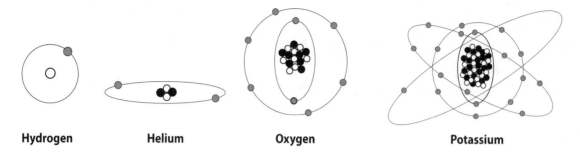

Hydrogen   Helium   Oxygen   Potassium

WHAT DID WE LEARN?

- What is an atom? **The smallest part of matter that cannot be broken down by ordinary chemical means.**

- What are the three parts of an atom? **Protons, neutrons, and electrons**

- What electrical charge does each part of the atom have? **Protons are positive, neutrons are neutral, and electrons are negative.**

- What is the nucleus of an atom? **The dense center of the atom consisting of protons and neutrons.**

- What part of the atom determines what type of atom it is? **The number of protons in the nucleus determines what kind of atom it is.**

- What is a valence electron? **An electron in the outermost energy level for that atom.**

TAKING IT FURTHER

- Why is it necessary to use a model to show what an atom is like? **Atoms are too small to see and are very complex, so a model is useful for understanding what an atom is like.**

- On your worksheet, you colored neutrons blue and protons red. Are neutrons actually blue and protons actually red in a real atom? **No, the colors used in a model are just to help us visualize the parts. They do not really represent the actual colors.**

CHALLENGE: ENERGY LEVELS WORKSHEET

Element	Energy levels	Electrons in level 1	Electrons in level 2	Electrons in level 3	Electrons in level 4	Electrons in level 5	Electrons in level 6
He Helium	1	2					
Be Beryllium	2	2	2				
Al Aluminum	3	2	8	3			
Cl Chlorine	3	2	8	7			
Fe Iron	4	2	8	14	2		
Kr Krypton	4	2	8	18	8		
Ag Silver	5	2	8	18	18	1	
Au Gold	6	2	8	18	32	18	1

ATOMIC MASS

HOW BIG IS AN ATOM?

SUPPLY LIST

Copy of "Learning About Atoms" worksheet
Supplies for Challenge: Copy of "Understanding Atoms" worksheet

LEARNING ABOUT ATOMS WORKSHEET

Element	Atomic number	Atomic mass	# of of protons	# of electrons	# of neutrons
Hydrogen	1	1	**1**	**1**	**0**
Helium	2	4	**2**	**2**	**2**
Oxygen	8	16	**8**	**8**	**8**
Fluorine	9	19	**9**	**9**	**10**
Chromium	24	52	**24**	**24**	**28**

WHAT DID WE LEARN?

- What are the three particles that make up an atom? **Proton, electron, and neutron.**
- What is the atomic number of an atom? **The number of protons in the nucleus.**
- What is the atomic mass of an atom? **The sum of the protons and neutrons in the nucleus of the atom.**
- How can you determine the number of electrons, protons, and neutrons in an atom if you are given the atomic number and atomic mass? **The number of protons is the same as the atomic number. The number of electrons is equal to the number of protons. The number of neutrons is equal to the atomic mass minus the number of protons.**

TAKING IT FURTHER

- What does a hydrogen atom become if it loses its electron? **A proton.**
- Why are electrons ignored when calculating an element's mass? **The mass of an electron is so small compared to the mass of a proton or neutron that it does not make a significant difference.**

CHALLENGE: UNDERSTANDING ATOMS WORKSHEET

Element	Symbol	Atomic number	Atomic mass	# of protons	# of electrons	Most common # of neutrons
Hydrogen	**H**	**1**	**1.008**	**1**	**1**	**0**
Oxygen	**O**	**8**	**16**	**8**	**8**	**8**
Boron	**B**	**5**	**10.81**	**5**	**5**	**6**
Gold	**Au**	**79**	**197**	**79**	**79**	**118**
Silver	**Ag**	**47**	**107.9**	**47**	**47**	**61**
Uranium	**U**	**92**	**238**	**92**	**92**	**146**

Potassium	**K**	19	39.1	19	19	20
Chlorine	**Cl**	17	35.45	17	17	18
Neon	**Ne**	10	20.18	10	10	10
Einsteinium	**Es**	99	252	99	99	153

LESSON 4

MOLECULES

PUTTING ATOMS TOGETHER

SUPPLY LIST

Copy of "What Am I?" worksheet

Supplies for Challenge: Copy of "Molecule Puzzle Pieces"   Scissors

WHAT AM I? WORKSHEET

Next to each of the substances below, write whether it is an element, a diatomic molecule or a compound. Review these terms in the lesson if you need to.

Gold (Au): **Element**

Oxygen (O_2): **Diatomic molecule**

Silver (Ag): **Element**

Ammonia (NH_3): **Compound**

Nitrogen (N_2): **Diatomic molecule**

Salt (NaCl): **Compound**

WHAT DID WE LEARN?

- What is a molecule? **Two or more atoms chemically connected or bonded together.**

- What is a diatomic molecule? **A molecule with two of the same type of atoms connected together.**

- What is a compound? **A molecule made from two or more different kinds of atoms.**

TAKING IT FURTHER

- What is the most important factor in determining if two atoms will bond with each other? **The number of valence electrons each atom has.**

- Table salt is a compound formed from sodium and chlorine. Would you expect sodium atoms and chlorine atoms to taste salty? Why or why not? **No, because when molecules are formed, the resulting compound is a new substance with its own characteristics, completely different from those of the original elements.**

ATOMS & MOLECULES

LESSONS 1–4

Label the parts of this helium atom.

A. _Electron_

B. _Proton or neutron_

C. _Neutron or proton_

B and C together form the _nucleus_.

Match the term with its definition.

1. _D_ Anything that has mass and takes up space

2. _E_ A positively charged particle in an atom

3. _I_ A negatively charged particle in an atom

4. _B_ A neutral particle in the nucleus

5. _A_ Mass of a proton or neutron

6. _J_ Compact center of the atom

7. _G_ Two atoms of the same element connected together

8. _F_ Part of matter that cannot be broken down chemically

9. _H_ Number of protons an element has

10. _C_ Two or more atoms chemically bonded

CHALLENGE QUESTIONS

11. Refer to the periodic table of the elements to complete the following chart.

Element	Atomic #	Atomic mass	# of protons	# of electrons	# of neutrons
Carbon	6	12.01	6	6	6
Aluminum	13	26.98	13	13	14
Tungsten	74	183.9	74	74	110

12. Based on the electron configurations for the following elements, which would not be likely to bond with any other elements? **Argon (Ar).**

13. What is an isotope? **An atom having the same number of protons but a different number of neutrons.**

14. What is a valence electron? **An electron in the outer shell of an atom.**

15. What is the electron configuration for silicon? **2, 8, 4.**

ELEMENTS

THE PERIODIC TABLE OF THE ELEMENTS

ORGANIZING THE ELEMENTS

SUPPLY LIST

Copy of "Learning about the Elements" worksheet

LEARNING ABOUT THE ELEMENTS WORKSHEET

1. What is the symbol for calcium? **Ca.**
2. What is the symbol for silver? **Ag.**
3. What is the atomic number for copper? **29.**
4. What is the atomic mass for rutherfordium? **261.**
5. What are two elements in the same column as sodium? **H, Li, K, Rb, Cs, Fr.**
6. What are two elements with eight electrons in their outer layer? **Ne, Ar, Kr, Xe, Rn.**
7. How many electrons are in the outer layer of nitrogen? **5.**
8. How many layers of electrons does barium have? **6.**
9. Name one transition element. **Accept anything from columns IB–VIIB.**
10. Would silicon be more likely to react the same way as carbon or chlorine? **Carbon—they are in the same column and have the same number of valence electrons.**

WHAT DID WE LEARN?

- How many valence electrons do the elements in each column have? **Those in column IA have 1 valence electron; those in column IIA have 2, IIIA have 3, etc. Those in columns IB–VIIIB have 1 or 2 valence electrons.**

- What four pieces of information are included for each element in any periodic table of the elements? **The element name, symbol, atomic number, and atomic mass.**

- What do all elements in a column on the periodic table have in common? **They have the same number of valence electrons.**

- What do all elements in a row on the periodic table have in common? **They have electrons in the same number of energy levels/same number of electron layers.**

TAKING IT FURTHER

- Atoms are stable when they have eight electrons in their outermost energy level. Therefore elements from column IA will react easily with elements from which column? **Column VIIA.**

- Elements from column IIA will react easily with elements from which column? **Column VIA.**

LESSON 6

METALS

SILVER AND GOLD HAVE I NONE . . .

SUPPLY LIST

Flashlight with battery   Electrical or duct tape   Copper wire

Supplies for Challenge: Copy of "Reactivity Series" worksheet

WHAT DID WE LEARN?

- What are the six characteristics of most metals? **Silvery luster, solid, malleable, ductile, conducts electricity, reacts with other elements.**

- How many valence electrons do most metals have? **Most commonly, metals have 1 or 2 valence electrons, but some have 3 or 4.**

- What is a metalloid? **An element that has some metal characteristics and some non-metal characteristics.**

TAKING IT FURTHER

- What are the most likely elements to be used in making computer chips? **The semiconductors—the ones shaded dark green on the periodic table of the elements. The most commonly used elements are silicon, germanium, and boron.**

- Is arsenic likely to be used as electrical wire in a house? **No, it is only a semi-conductor so it would not make good electrical wiring.**

CHALLENGE: REACTIVITY SERIES WORKSHEET

- **1. Sodium 2. Potassium 3. Magnesium 4. Calcium 5. Potassium 6. Lead 7. Zinc 8. Magnesium 9. Sodium 10. Calcium**

LESSON 7

NON-METALS

THE REST OF THE ELEMENTS

SUPPLY LIST

Eggs   Vinegar   Fluoride toothpaste   Nail polish or permanent marker

WHAT DID WE LEARN?

- What are some common characteristics of non-metals? **Not shiny or silver, not conductive, do not easily lose electrons.**

- What is the most common state, solid, liquid, or gas, for non-metal elements? **Gas.**

- Why are halogens very reactive? **They need only one electron to fill their outer shells.**

- Why are noble gases non-reactive? **They have a full outer shell of electrons.**

TAKING IT FURTHER

- Hydrogen often acts like a halogen. How might it act differently from a halogen? **Because hydrogen has only one electron, it can give up its electron and become an ion, whereas halogens do not easily give up electrons.**

- Why are balloons filled with helium instead of hydrogen? **Helium is a noble gas and non-reactive, but hydrogen is highly reactive. We don't want balloons exploding as the hydrogen reacts with another element.**

HYDROGEN

VERY REACTIVE

SUPPLY LIST

Vegetable oil   Margarine   Peanut butter   Prepackaged food labels

WHAT DID WE LEARN?

- What is the most common element in the universe? **Hydrogen.**
- What is the atomic structure of hydrogen? **It has one proton and one electron.**
- What is the atomic number for hydrogen? **1.**
- What is the most common element in the universe? **Hydrogen.**
- Why is hydrogen sometimes grouped with the alkali metals? **It has only one electron so it often behaves like an alkali metal.**
- Why is hydrogen sometimes grouped with the halogens? **It is stable if it gains one electron so it often behaves like a halogen.**

TAKING IT FURTHER

- Why is hydrogen one of the most reactive elements? **Most elements must either gain electrons or lose electrons to combine with other elements. But hydrogen can do either one so it combines easily with many other elements.**

- Margarine contains only partially hydrogenated oil. What do you suppose fully hydrogenated oils are like? **They are much harder or more solid than margarine and are not easily spread.**

CARBON

GRAPHITE AND DIAMONDS

SUPPLY LIST

Drawing paper   Colored pencils   Ceramic plate   Candle   Matches

WHAT DID WE LEARN?

- What is the atomic number and atomic structure of carbon? **Carbon is element number 6. It has 6 protons, 6 neutrons, and 6 electrons.**
- What makes a compound an organic compound? **It contains carbon atoms.**
- Name two common forms of carbon. **Graphite and diamond.**
- What is one by-product of burning coal? **Carbon dioxide.**

TAKING IT FURTHER

- How does the carbon cycle demonstrate God's care for His creation? **It allows carbon to be recycled and keeps life continuing on earth.**
- What is the most likely event that caused coal formation? **The Genesis Flood would have buried large amounts of plants under tons of mud and water. This is the most likely cause of the large amounts of coal found in the earth.**
- What would happen if bacteria and fungi did not convert carbon into carbon dioxide gas? **The carbon from dead plants and animals would become trapped and would not be able to be reused in the growth of new plants.**

LESSON 10

OXYGEN

A VERY ESSENTIAL ELEMENT

SUPPLY LIST

Candle   Small piece of dry ice   Matches   Gloves   Glass cup
Supplies for Challenge: Steel wool   2 test tubes   Small dish   Dish soap

WHAT DID WE LEARN?

- What is the atomic structure of oxygen? **Oxygen has 8 protons and 8 neutrons in the nucleus, and 8 electrons. It has 6 valence electrons.**
- How is ozone different from the oxygen we breathe? **Ozone is a molecule of three oxygen atoms. The oxygen we breathe is a molecule of two oxygen atoms. O_3 is poisonous and O_2 is not.**

TAKING IT FURTHER

- Why does the existence of ozone in the upper atmosphere show God's provision for life on earth? **If ozone were in the lower atmosphere, it would poison all living things. But in the upper atmosphere, it protects the earth from harmful radiation.**
- How do animals in the ocean get the needed oxygen to "burn" the food they eat? **Most aquatic animals have gills that extract oxygen from the water. A few, like whales and dolphins, have to surface and breathe air.**
- Why are oxygen atoms nearly always combined with other atoms? **They have only six valence electrons, so they are not stable by themselves.**

ELEMENTS

LESSONS 5–10

Short answer:

1. What do elements in a column of the periodic table have in common? **The same number of valence electrons.**

2. What do elements in a row of the periodic table have in common? **Electrons filling the same energy levels.**

3. Which column of elements is most stable? **VIIIA.**

4. Elements in which column are most likely to react with elements in column VIA? **IIA.**

5. Elements in which column are most likely to react with elements in column VIIA? **IA.**

Write metal, metalloid, or non-metal to match the type of element to its characteristics.

6. _**Metal**_ Silvery luster

7. _**Metal**_ Ductile

8. _**Metal**_ Conducts electricity

9. _**Non-metal**_ Does not conduct electricity

10. _**Metal**_ Solid at room temperature

11. _**Non-metal**_ Not shiny

12. _**Metalloid**_ Somewhat malleable

13. _**Non-metal**_ Most often a gas

14. _**Metalloid**_ Semiconductor

15. _**Metal**_ Malleable

Mark each statement as either True or False.

16. _**T**_ Hydrogen is very reactive.

17. _**F**_ Oxygen is lighter than hydrogen.

18. _**T**_ Hydrogen is sometimes grouped with alkali metals.

19. _**T**_ Hydrogen is sometimes grouped with halogens.

20. _**F**_ Hydrogen is the most common element on earth.

21. _**T**_ All elements are recycled—they are not destroyed.

22. _**T**_ Carbon forms organic compounds.

CHALLENGE QUESTIONS

Match the term to its definition.

23. _**B**_ Column of the periodic table

24. _**F**_ Row of the periodic table

25. _**A**_ Metals in column 1

26. _**D**_ Metals in column 2

27. _**E**_ Non-reactive metals

28. _**G**_ Ball-shaped carbon molecule

29. _**H**_ Thread-like cylinders of carbon atoms

30. _**C**_ Technology that combines hydrogen and oxygen to produce electricity

UNIT 3
BONDING

LESSON 11

IONIC BONDING

GIVING UP ELECTRONS

SUPPLY LIST

Colored mini-marshmallows   Toothpicks   Glue

Supplies for Challenge: Copy of "Name that Ion" worksheet

WHAT DID WE LEARN?

- What is the main feature in an atom that determines how it will bond with other atoms? **The number of valence electrons it contains.**

- What kind of bond is formed when one atom gives up electrons and the other atom takes the electrons from it? **An ionic bond.**

- What is electronegativity? **A measure of how tightly an element holds on to its valence electrons.**

- Why are compounds that are formed when one element takes electrons from another called ionic compounds? **Because ions are formed when electrons are taken away or added.**

- What are some common characteristics of ionic compounds? **Conduct electricity when melted or dissolved, high melting point, soluble in water, brittle, form ions, form crystal lattices.**

- Which element has a higher electronegativity, chlorine or potassium? **Electronegativity increases as you go from left to right across the periodic table. Chlorine holds on to its electrons more tightly than potassium, so it has a higher electronegativity.**

TAKING IT FURTHER

- Which column of elements are the atoms in column IA most likely to form ionic bonds with? **The elements in column VIIA.**

- Use the periodic table of the elements to determine the number of electrons that barium would give up in an ionic bond. **Barium has 2 valence electrons that it would give up.**

CHALLENGE: NAME THAT ION WORKSHEET

NaF: **Sodium fluoride**   KCl: **Potassium chloride** $CaCl_2$: **Calcium chloride or calcium dichloride**

LiBr: **Lithium bromide**   CaS: **Calcium sulfide**

COVALENT BONDING

SHARING ELECTRONS

SUPPLY LIST

Colored mini-marshmallows   Toothpicks   Glue

Supplies for Challenge: Distilled water   9-volt battery   Copper wire   Sugar   Salt

Baking soda   Olive oil   4 paper cups   Copy of "Bonding Experiment" worksheet

WHAT DID WE LEARN?

- What is a covalent bond? **A bond formed when electrons are shared between two or more atoms.**

- What are some common characteristics of covalent compounds? **Do not conduct electricity, low melting point, strong, flexible, lightweight, insoluble in water, only slight attraction for each other.**

- What is the most common covalent compound on earth? **Water.**

TAKING IT FURTHER

- Why do diatomic molecules form covalent bonds instead of ionic bonds? **Diatomic molecules are formed from two atoms of the same element, so they have the same electronegativity. Since neither atom is able to take away or give up its electrons, they cannot form ionic bonds.**

- Would you expect more compounds to form ionic bonds or covalent bonds? **Since there are so many metals and only a few metalloids and non-metals, you might expect most compounds to be ionic. However, there are actually so many different ways to share electrons that covalent bonds are actually more common.**

CHALLENGE: BONDING EXPERIMENT WORKSHEET

- Use a periodic table to determine if each of the following compounds is composed of metals, non-metals, or both.

Water—H_2O is composed of _**non-metals**_

Baking Soda—$NaHCO_3$ is composed of _**both**_

Sugar (sucrose)—$C_{12}H_{22}O_{11}$ is composed of _**non-metals**_

Salt—$NaCl$ is composed of _**both**_

Olive oil—$C_{17}H_{35}COOH$ is composed of _**non-metals**_

Compound tested	Will it conduct electricity? (Hypothesis)	Did it conduct electricity? (Observations)	Ionic or covalent? (Conclusions)
Distilled water		No	Covalent
Baking soda		Yes	Ionic
Sugar		No	Covalent
Salt		Yes	Ionic
Olive oil		No	Covalent

LESSON 13

METALLIC BONDING

SHARING ON A LARGE SCALE

SUPPLY LIST

Colored mini-marshmallows   Toothpicks   Glue
Supplies for Challenge: Copy of "Bonding Characteristics" worksheet

WHAT DID WE LEARN?

- What is the free electron theory? **It is the theory that metals form bonds by sharing electrons on a very large scale. Thousands of atoms allow their electrons to freely move about so that the atoms remain stable.**
- How many valence electrons do metals usually have? **Usually 1, 2, or 3.**
- What are common characteristics of metallic compounds? **Free electrons, conduct electricity and heat, shiny luster, high melting point, insoluble in water.**

TAKING IT FURTHER

- Why don't metals form ionic or covalent bonds? **Because they have similar numbers of valence electrons, they do not pull electrons away from each other. Also, because they have a low number of valence electrons, they do not have enough to share among a small number of atoms. Therefore, they must share on a large scale—among thousands of atoms.**
- Would you expect semiconductors to form metallic bonds? **No. Since they do not conduct electricity well, they would not have free electrons.**

CHALLENGE: BONDING CHARACTERISTICS WORKSHEET

Ionic bonding	Covalent bonding	Metallic bonding
A	C	B
F	E	D
G	H	I
J	K	J
L	M	L
O	N	N

LESSON 14

MINING & METAL ALLOYS

MAKING IT STRONGER

SUPPLY LIST

Tarnished silver object   Silver polish   Soft cloth
Supplies for Challenge: Copy of "Common Alloys" worksheet

What did we learn?

- What element is combined with most metals to form metal ore? **Most metals are in the form of metal oxides—metals combined with oxygen.**

- What must be done to metal oxides to obtain pure metal? **The oxygen must be removed through a reduction reaction.**

- What is an alloy? **A metal that has a small amount of another metal added to it.**

- Why are alloys produced? **Alloys are often stronger, more resilient, and easier to work with than pure metals.**

Taking it further

- Do you think chromium would be added to steel that is going to be used in saw blades? Why or why not? **Probably not. Chromium keeps steel from oxidizing; however, a little oxidation on a saw blade will not keep it from working. The saw blade needs to be strong, so tungsten may be added, but not chromium.**

- Is oxidation of metal always a bad thing? **Not always. Sometimes a layer of oxidation prevents more oxygen from reaching the rest of the metal. So leaving a small amount of oxidation can actually reduce the overall amount of oxidation that occurs. This is not always the case, however; sometimes oxidation, such as rust, continues to occur until the sample is completely gone.**

Challenge: Common Alloys worksheet

Bronze: **Copper, tin**

Brass: **Copper, zinc**

Steel: **Iron, carbon**

Solder: **Tin, lead**

Duraluminium: **Aluminum, copper**

Magnalium: **Aluminum, magnesium**

Pewter: **Tin, copper, sometimes antimony or bismuth**

Sterling silver: **Silver and usually copper**

Stainless steel: **Iron, carbon, nickel, chromium**

LESSON 15

CRYSTALS

Sparkling like diamonds

Supply list

Table salt   2 plates   Epsom salt   Scissors   Dark construction paper   Small pan
Optional activity: geode

Supplies for Challenge: Plaster of Paris   Modeling clay

What did we learn?

- What is a crystal? **A substance whose atoms are lined up in a regular lattice configuration. Crystals have smooth faces and defined edges.**

- How do crystals form? **When a liquid cools slowly, the atoms line up in regular patterns to form crystal lattices based on their chemical characteristics.**

- What is an artificial gem? **One that is formed by man and not formed naturally.**

- Where would you look to find crystals? **In rocks and minerals, in the kitchen (salt and sugar), in caves, jewelry.**

TAKING IT FURTHER

- Why are naturally occurring gems more valuable than artificial gems when many are made from the same materials? **Even though they are made from the same materials, artificial gems do not have the same strength and brilliance of naturally occurring crystals. God's crystals are still better than man's.**

- Why is a saturated solution better for forming crystals? **The more atoms of the crystal forming material you have, such as salt, the more likely they are to line up in a lattice formation.**

- What are some ways you use crystals in your home? **In food, in your computer, TV, phone, and other electronic devices, in your rock collection, gems in your mother's wedding ring, etc.**

LESSON 16

CERAMICS

MAKING IT WITH CLAY

SUPPLY LIST

Polymer clay (Femo, Sculpey, etc.)

WHAT DID WE LEARN?

- What is ceramic? **It is a material that is formed when ingredients fuse together by heat; often made with clay.**
- What are some examples of traditional ceramics? **Pottery, brick, porcelain, and glass.**
- What makes ceramics hard? **The material forms crystals when it is baked or fired.**
- What are some advantages of modern ceramics? **They are hard, strong, heat resistant, and don't rust.**

TAKING IT FURTHER

- Why are the tiles on the space shuttle made of ceramic? **Because ceramic is very heat resistant, the tiles keep the heat generated by friction with the atmosphere away from the shuttle, allowing the shuttle to reenter the atmosphere without burning up.**

- Why are crystalline structures stronger than non-crystalline structures? **The lattice shape of the bonds allows atoms to be connected in more than one direction, so the compounds are stronger.**

QUIZ 3

BONDING

LESSONS 11–16

For each characteristic below, write I if it describes an ionic bond, C for a covalent bond, and M for a metallic bond. Some characteristics have more than one answer.

1. _I_ Formed by elements with very different levels of electronegativity
2. _I,M_ High melting point
3. _C_ Electrons are shared between two or three atoms
4. _C,M_ Insoluble in water

5. _I_ Forms ions

6. _I_ Electrons are given up or pulled away

7. _C_ Does not conduct electricity

8. _M_ Sharing of electrons on a large scale

9. _I,M_ Conducts electricity

10. _C_ Flexible

Short answer:

11. How are crystals formed? **When a liquid slowly cools the atoms may line up in specific patterns to form crystals.**

12. What is the smooth side of a crystal called? **A face.**

13. What process is necessary for ceramics to become strong? **Heating or firing.**

14. What is the common ingredient in all natural ceramics? **Clay.**

15. Name three traditional ceramics. **Pottery, brick, porcelain, glass.**

CHALLENGE QUESTIONS

Mark each statement as either True or False.

16. _T_ Ionic bonding occurs between elements with very different electronegativities.

17. _F_ Ions are electrically neutral.

18. _F_ Ionic bonds occur between non-metals

19. _T_ Sodium fluoride is an ionic compound.

20. _F_ Covalent compounds easily conduct electricity.

21. _T_ Covalent bonds occur between non-metals.

22. _F_ Metallic materials easily dissolve in water.

23. _T_ Metallic bonds have free electrons.

24. _T_ Metallic bonds form between elements with similar low electronegativities.

25. _T_ Brass is an alloy of copper and zinc.

26. _F_ Steel is an allow of copper and tin.

27. _F_ Hydrates usually feel wet.

28. _T_ Hydrates can help prevent the spread of fire.

29. _T_ Resorbable ceramics are absorbed into the body.

30. _F_ Inert ceramics react with the body.

CHEMICAL REACTIONS

LESSON 17

CHEMICAL REACTIONS

CHANGING FROM ONE THING TO ANOTHER

SUPPLY LIST

Birthday candle   Vinegar   Modeling clay   Baking soda   Jar   Matches
Supplies for Challenge: 6 clear cups   6 Alka-Seltzer tablets   Water   Ice   Spoon
Stopwatch   Copy of "Reaction Rate Experiment" worksheet

WHAT DID WE LEARN?

- What is a chemical reaction? **When atomic bonds are formed or broken—when two or more elements combine together to form a new substance, or when a substance is broken down into its separate elements.**

- What are the initial ingredients in a chemical reaction called? **Reactants.**

- What are the resulting substances of a chemical reaction called? **Products.**

TAKING IT FURTHER

- How might you speed up a chemical reaction? **Add heat; add surface area to the reactants by changing their shape—make them thinner or break or crush them; increase the concentration of the reactants; add a catalyst.**

- A fire hose usually sprays water on a fire to put it out. Water does not deprive the fire of oxygen, so why does water put out a fire? **Water absorbs the heat from the fire, and heat is another necessary ingredient in producing and sustaining a fire.**

- What chemical reaction do you think is taking place in the making of a loaf of bread? **The yeast reacts with the sugar in the bread dough to produce carbon dioxide.**

CHALLENGE: REACTION RATE EXPERIMENT WORKSHEET

- **You should see the tablet in the hot water dissolve more quickly than the tablets in the other cups. You should see the crushed tablet dissolve more quickly than th tablets in the other cups.**

CHEMICAL EQUATIONS

DESCRIBING HOW IT WORKS

SUPPLY LIST

Copy of "Understanding Chemical Equations" worksheet

Supplies for Challenge: Copy of "Reactants and Products" worksheet

UNDERSTANDING CHEMICAL EQUATIONS WORKSHEET

1. $C + O_2 \longrightarrow CO_2$
2. $N_2 + 3H_2 \longrightarrow 2NH_3$
3. $2H_2O \longrightarrow 2H_2 + O_2$

WHAT DID WE LEARN?

- What is a chemical equation? **It is an equation that visually shows what happens to each element in a chemical reaction.**

- What are the elements or compounds on the left side of a chemical equation called? **The reactants.**

- What are the elements or compounds on the right side of a chemical equation called? **The products.**

TAKING IT FURTHER

- Why is it helpful to use chemical equations? **Equations provide a visual way to see what is happening in a chemical reaction without drawing pictures.**

CHALLENGE: REACTANTS & PRODUCTS WORKSHEET

1. $4\,Al + 3O_2 \longrightarrow$ _**B**_
2. $H_2SO_4 + 2\,LiOH \longrightarrow$ _**A**_
3. $4\,NH_3 + 3\,O_2 \longrightarrow$ _**C**_
4. $P_4 + 10\,Cl_2 \longrightarrow$ _**E**_
5. $CO_2 \longrightarrow$ _**D**_
6. $H + OH \longrightarrow$ _**F**_
7. $2\,KClO_3 \longrightarrow$ _**H**_
8. $2\,Na + 2\,H_2O \longrightarrow$ _**G**_

- Which of the above equations represent decomposition reactions? _**5, 7**_
- Which of the above equations represent composition reactions? _**1, 4, 6**_
- Which of the above equations represent single displacement reactions? _**3, 8**_
- Which of the above equations represent double displacement reactions? _**2**_

LESSON 19

CATALYSTS

SPEEDING THINGS UP

SUPPLY LIST

Potato   Hydrogen peroxide   Apple   Lemon juice

WHAT DID WE LEARN?

- What is a catalyst? **A substance added to speed up a chemical reaction.**
- How does a catalyst work? **It reduces the amount of energy needed for the chemical reaction to take place.**
- What is an inhibitor? **A substance that slows down or prevents a chemical reaction.**
- What is an enzyme? **A catalyst found in living cells.**

TAKING IT FURTHER

- Why is it important that living cells have enzymes? **If enzymes were not available, many chemical reactions such as digestion would take much too long to occur.**
- Are catalysts always good? **Not necessarily. If a catalyst caused food to spoil very quickly that would be a bad use of a catalyst.**

LESSON 20

ENDOTHERMIC & EXOTHERMIC REACTIONS

WHAT HAPPENS TO THE HEAT?

SUPPLY LIST

5 eggs   Vinegar (room temperature)   Small pan   Thermometer   Timer
Steel wool (no soap)   Jar with lid (Thermometer must fit inside the jar with the lid on)
Supplies for Challenge: Alka-Seltzer tablets   Thermometer   Styrofoam cup
Water   Copy of "Endothermic or Exothermic?" worksheet   Stopwatch

WHAT DID WE LEARN?

- What is an exothermic reaction? **A chemical reaction that releases energy.**
- What is an endothermic reaction? **A chemical reaction that absorbs energy.**

TAKING IT FURTHER

- If a chemical reaction produces a spark, is it likely to be an endothermic or exothermic reaction? **Light is a form of energy, so it would be an exothermic reaction.**
- How does photosynthesis and digestion reveal God's plan for life? **Photosynthesis absorbs and stores energy from the sun in the sugar molecules in the plant. That energy is released during digestion after an animal eats the plant. This is God's plan for providing necessary food, and therefore energy, for all of the animals—and humans—on earth.**

- If the temperature of the product is lower than the temperature of the reactants, was the reaction endothermic or exothermic? **If the result is cooler than the beginning reactants, then energy was absorbed, so the reaction was endothermic.**

CHALLENGE: ENDOTHERMIC OR EXOTHERMIC? WORKSHEET

- **The results should show the reaction is endothermic; the temperature of the water goes down during the reaction then levels off.**

QUIZ 4 CHEMICAL REACTIONS

LESSONS 17–20

Mark each statement as either True or False.

1. _F_ All chemical reactions are fast.
2. _T_ A catalyst speeds up a chemical reaction.
3. _T_ Endothermic reactions use up heat.
4. _F_ A fireworks explosion is an endothermic reaction.
5. _T_ The same number of atoms must appear on both sides of a chemical equation.
6. _T_ Chemical equations demonstrate the first law of thermodynamics.
7. _T_ Reactants are on the left side of a chemical equation.
8. _F_ Catalysts are used up in a chemical reaction.
9. _F_ Inhibitors speed up a chemical reaction.
10. _T_ Sometimes inhibitors are helpful.
11. _T_ Catalysts lower the energy required for a chemical reaction to occur.
12. _T_ Exothermic reactions release energy.
13. _F_ The product of an exothermic reaction is cooler than the reactants.
14. _F_ Chemical reactions are rare.
15. _T_ Heat can increase the reaction rate.

CHALLENGE QUESTIONS

16. _Double displacement_ $H_2SO_4 + 2\,LiOH \longrightarrow Li_2SO_4 + 2\,H_2O$
17. _Composition_ $P_4 + 10\,Cl_2 \longrightarrow 4\,PCl_5$
18. _Decomposition_ $CO_2 \longrightarrow C + O_2$
19. _Single displacement_ $2\,Na + 2\,H_2O \longrightarrow H_2 + 2\,NaOH$
20. _Double displacement_ $AgNo_3 + HCl \longrightarrow AgCl + HNO_3$

Short answer:

21. List three ways to increase the reaction rate of a chemical reaction: **Add heat, increase surface area of reactants, increase concentration of reactants, add a catalyst, reduce activation energy.**

22. List two groups of catalysts. **Heterogeneous, homogeneous.**

23. Which type of catalyst is found in a catalytic converter? **Heterogeneous.**

24. Which type of catalyst will bond with a reactant? **Homogeneous.**

25. What is the name for the energy stored in chemical bonds? **Enthalpy.**

UNIT 5
ACIDS & BASES

LESSON 21

CHEMICAL ANALYSIS
WHAT IS IT MADE OF?

SUPPLY LIST

Red or purple cabbage   Pan or microwavable bowl

Supplies for Challenge: Research materials on chemical analysis

WHAT DID WE LEARN?

- What is chemical analysis? **Using chemical reactions to determine the composition of a substance.**
- List three different types of chemical analysis. **Flame test, spectrometer, indicators.**
- What is a chemical indicator? **A substance that changes color when it reacts with a specific chemical.**
- What is the pH scale? **The scale used to measure the strength of an acid or a base.**
- What does a pH of 7 tell you about the substance? **It is neutral. It is not an acid or a base.**

TAKING IT FURTHER

- Why is it important to periodically test the pH of swimming pool water? **Water must be close to neutral to be safe to swim in. Also, water with a pH much greater than 6.8–7.0 can cause pipes to become clogged with minerals.**
- Name at least one other use for testing pH of a liquid? **Hair treatments like permanents must be tested for pH so that hair curls and doesn't burn. Urine can be tested for pH to detect health problems. Beverages are tested for proper pH to ensure proper taste. Drinking water is tested for proper pH, and wastewater is tested before releasing it back into the water system.**

LESSON 22

ACIDS
DOES IT BURN?

SUPPLY LIST

Lemon juice   Vinegar   Lemon lime soda   Milk   Cabbage indicator from lesson 21

Supplies for Challenge: Jar with a lid   15 pennies   1 steel paper clip   Salt   Vinegar

TESTING FOR ACIDS

- **Lemon juice, vinegar, and soda pop should be acids. Saliva should be slightly acidic; milk should not be an acid.**

WHAT DID WE LEARN?

- What defines a substance as an acid? **It produces hydronium ions when dissolved in water.**
- What is a hydronium ion? H_3O^+, **formed by a water molecule and a hydrogen ion.**
- How is a weak acid different from a strong acid? **A weak acid holds onto its hydrogen atoms more strongly than a strong acid, so it forms fewer hydronium ions in water.**
- What are some common characteristics of an acid? **Sour taste, conducts electricity in water, reacts with metals, many are corrosive, neutralizes bases, reacts with indicators.**
- How can you tell if a substance is an acid? **Dissolve it in water and use an indicator to test for acid.**

TAKING IT FURTHER

- Why is saliva slightly acidic? **The acid in your saliva helps begin the digestion process by helping break down the food molecules.**
- Would you expect water taken from a puddle on the forest floor to be acidic, neutral, or basic? Why? **It would probably be acidic because the forest floor is covered with decaying plants, and decaying plants produce humic acid.**
- What would you expect to be a key ingredient in sour candy? **Some kind of acid. Sour spray and other sour candies often contain several types of acids.**

LESSON 23

BASES

THE OPPOSITE OF ACIDS

SUPPLY LIST

Ammonia (clear)   Soap   Anti-acid tablets or liquid   Baking soda   Toothpaste
Cabbage indicator from lesson 21

Supplies for Challenge: Ammonia   Eyedropper   Distilled water   Measuring cup
Measuring spoon   Clear glass   Vinegar   Cabbage indicator

TESTING FOR BASES

- **They should all be bases. However, if you tested liquid soap and the indicator showed it to be an acid or to be neutral, check the bottle to see if it contains some kind of citric acid, which is sometimes added for scent.**

WHAT DID WE LEARN?

- What defines a substance as a base? **It produces hydroxide ions when dissolved in water.**
- What is a hydroxide ion? **OH⁻ ion.**
- How is a weak base different from a strong base? **A weak base holds onto its hydroxide ions more strongly than a strong base does.**
- What are some common characteristics of a base? **Bitter taste, conducts electricity in water, feels slippery, many are corrosive, neutralizes acids, reacts with indicators.**

- How can you tell if a substance is a base? **Dissolve it in water and use an indicator to test for base.**

TAKING IT FURTHER

- If you spill a base, what should you do before trying to clean it up? **Add an acid to neutralize it.**
- Do you think that Strontium is likely to form a strong base? Why or why not? **Strontium is in the alkali metal family and alkali metals tend to form strong bases. Therefore, strontium is likely to form a strong base.**

SALTS

PASS THE SALT, PLEASE.

SUPPLY LIST

Lemon juice   Anti-acid (tablet or liquid)   Table salt   Swabs
Supplies for Challenge: Copy of "Acid/Base Reactions" worksheet

WHAT DID WE LEARN?

- Did you detect the various flavors in the areas indicated on the above tongue map? **Answers may vary.**
- How is a salt formed? **When a negative acid ion combines with a positive base ion, a salt is formed.**
- What are two common characteristics of salts? **They have a salty flavor, and they form crystals.**
- How are salt families named? **By the acid from which they are made.**
- Name three salt families. **Sulfates, chlorides, nitrates, carbonates, phosphates, potash.**

TAKING IT FURTHER

- What do you expect to be the results of combining vinegar and lye? **You would get a salt and water.**
- Why are some salts still acidic or basic? **The ions do not completely combine together, so some hydrogen or hydroxide ions are still present.**
- If your tongue can only detect four different flavors, how can foods and drinks have so many different flavors? **There are many different combinations of acids, bases, salts, and sugars in foods so there is a great variety. Also, flavor is not just what you taste on your tongue. It also includes the smell of the food as well.**

CHALLENGE: ACID/BASE REACTIONS WORKSHEET

1. $HClO_3 + KOH \longrightarrow KClO_3 + H_2O$
   The acid is **_HClO3_**   The base is **_KOH_**   The salt is **_KClO$_3$_**
2. $HBr + Ca(OH)_2 \longrightarrow CaBr_2 + H_2O$
   The acid is **_HBr _**   The base is **_Ca(OH)$_2$_**   The salt is **_CaBr$_2$_**
3. $H_2SO_4 + 2NH_3 \longrightarrow 2NH_4^+ + SO_4^{2-}$
   The acid is **_H$_2$SO$_4$_**   The base is **_NH$_3$_**
4. $HI + H_2O \longrightarrow H_3O^+ + I^-$
   The acid is **_HI_**   The base is **_H$_2$O_**

ACIDS & BASES

LESSONS 21–24

Choose the best answer for each question.

1. _D_ Which is not a type of chemical analysis?
2. _A_ pH indicators can tell the strength of which type of compound?
3. _B_ What flower can indicate the pH of the soil by the color of its flowers?
4. _C_ Which of the following is not an acid?
5. _A_ Which of the following is not a base?
6. _A_ What is formed when an acid combines with a base?
7. _B_ Which is not a characteristic of acids?
8. _A_ Which is not a characteristic of bases?
9. _A_ Which acid is the most produced chemical in the United States?
10. _C_ What common product is made primarily from salts?

CHALLENGE QUESTIONS

Choose the best answer for each statement.

11. _C_ Electroplating is depositing a thin layer of metal on a **_conductor_**.
12. _A_ Titration allows you to calculate how many **_molecules_** are in an unknown sample.
13. _B_ A proton donor is another name for a(n) **_acid_**.
14. _C_ A proton acceptor is another name for a(n) **_base_**.
15. _D_ You can identify the acid in an equation because it loses a(n) **_hydrogen_** atom.

UNIT 6
BIOCHEMISTRY

LESSON 25

BIOCHEMISTRY

THE CHEMISTRY OF LIFE

SUPPLY LIST

Whatever food you have in your kitchen

Supplies for Challenge: Box of gelatin mix   Fresh pineapple juice   Vinegar   4 marbles
4 tall narrow cups   Measuring spoon   Copy of "Enzyme Reaction Rates" worksheet

WHAT DID WE LEARN?

- List at least two chemical functions performed inside living creatures. **Photosynthesis and digestion, or oxygen combining with hemoglobin. In plants, the equivalent of digestion is called internal or cellular respiration.**

- What is the chemical reaction that takes place during photosynthesis? **Water and carbon dioxide chemically combine to form sugar and oxygen.**

- What is the main chemical reaction that takes place during digestion? **Sugar and oxygen chemically combine to form carbon dioxide and water; also larger molecules are broken down into smaller molecules.**

- What substance is necessary for nearly every chemical reaction in living things? **Water.**

- Name the three major chemicals your body needs that are found in the foods we eat. **Proteins, fats, and carbohydrates.**

TAKING IT FURTHER

- Why did God design your body to have enzymes? **Enzymes help digestion and other metabolic processes to occur at a much quicker rate than they otherwise would.**

- With what you know about chemical processes, why do you think it is important to brush your teeth after you eat? **The chemicals in your mouth begin the digestion process. These chemicals can cause tooth decay if they stay in your mouth too long. So you need to brush away any food and acids so your teeth stay healthy.**

- Can you think of other chemical processes in your body besides the ones mentioned in this lesson? **Chemicals called hormones control your growth; chemicals are released to make you feel sleepy at bedtime; taste and smell are chemical reactions. These are just a few examples. The list of chemical reactions in your body is very long!**

CHALLENGE: ENZYME REACTION RATES WORKSHEET

1. What effects did the plain pineapple juice have on the gelatin? **The plain pineapple juice should break down the protein in the gelatin, allowing the marble to move down into the gelatin.**

2. What effects did heating the juice have on the way the juice affected the gelatin? **The heated juice should be less effective, allowing the marble to move more slowly.**

3. What effects did changing the pH have on the way the juice affected the gelatin? **Similarly, the vinegar changes the pH and makes the protease less effective, so the marble moves more slowly.**

4. Why was cup 4 necessary? **Cup 4 is a control; it shows that the movement of the marbles is due to the juice that was added and not just due to gravity.**

LESSON 26

DECOMPOSERS

ULTIMATE RECYCLING

SUPPLY LIST

Paper   Colored pencils

Supplies for Challenge: Banana   3 Plastic zipper bags   Baking yeast   Marker

Copy of "Rate of Decomposition" worksheet

WHAT DID WE LEARN?

- What is a scavenger? **An animal that eats dead animals.**

- What is a decomposer? **An organism that breaks down dead plants, dead animals or dung into simple chemical compounds.**

- What is this way of recycling nitrogen called? **The nitrogen cycle.**

TAKING IT FURTHER

- Why are decomposers necessary? **They are needed to break down complex compounds into simple compounds that can be used by plants. Without decomposers, the elements would be locked up and plants would not be able to grow.**

- Were there animal scavengers in God's perfect creation, before the Fall of man? **No, there was no animal or human death before Adam sinned. Man and animals were all created to be vegetarians—see Genesis 1:27-31.**

- Explain how a compost pile allows you to participate in the nitrogen cycle. **You can take food scraps, such as potato peels, and place them in a bin or pile outside. Bacteria or other decomposers eat these scraps, leaving behind compost, which is nutrient rich material that you can add to your garden. You have taken nitrogen from the food scraps and returned it to the soil to be used by the plants you grow in your garden.**

LESSON 27

CHEMICALS IN FARMING

HELPING PLANTS GROW

SUPPLY LIST

2 identical plants—a fast growing plant like mint is a good choice   Plant food or fertilizer

WHAT DID WE LEARN?

- What are three ways that farmers ensure their soil will have enough nutrients for their crops? **Adding fertilizers, allowing the land to lie fallow, crop rotation, burning of unwanted plants.**

- What is hydroponics? **Growing plants without soil, using chemicals in water.**

- How are chemicals used in farming other than for nutrients for the plants? **Chemicals are used to kill pests, diseases, and unwanted plants—pesticides, fungicides, and herbicides.**

- How is an organic farm different from other farms? **Organic farms do not use man-made chemicals.**

TAKING IT FURTHER

- Why did the farmers let cattle graze on their land once every fourth year in the Norfolk 4-course plant rotation method? **The animal waste added nutrients back into the soil.**

- How does hydroponics replace the role of soil in plant growth? **A framework is provided to support the plants, and nutrients are added to the water for absorption by the roots.**

LESSON 28 MEDICINES

HOW CHEMICAL COMPOUNDS AFFECT YOUR BODY

SUPPLY LIST

Bread   Butter or margarine   Garlic powder   Ginger ale

WHAT DID WE LEARN?

- Why are chemicals used as medicines? **Your body is constantly performing chemical reactions, so adding chemicals to your body causes different reactions to occur.**

- What were the earliest recorded medicines? **Herbs.**

- What was Sir Alexander Fleming's important discovery? **Penicillin—the first antibiotic.**

TAKING IT FURTHER

- If plants have the potential of supplying new medicines, where might a person look to find different plants? **One of the likeliest sources of medicinal plants is believed to be the tropical rainforests where there are thousands of unusual plants.**

- What other sources might there be for discovering new medicines? **In addition to plants, animals in the rainforest and ocean are likely places to test for new medicines. Also, a better understanding of how the human body processes chemicals can lead to the development of new synthetic medicines.**

QUIZ 6

BIOCHEMISTRY

LESSONS 25–28

Short answer:

1. Identify two chemical reactions that sustain life. **Photosynthesis, digestion.**
2. Name three main chemical compounds in food. **Carbohydrates, proteins, fat.**
3. What type of catalyst increases the rate of digestion processes? **Enzymes.**
4. What is an animal called that eats dead animals? **Decomposer or scavenger.**
5. Name two types of decomposers. **Bacteria, fungi.**
6. Name an element that is recycled by decomposers. **Nitrogen.**
7. Name three ways to keep farmland productive. **Crop rotation, lying fallow, fertilizers.**
8. Who was the discoverer of penicillin? **Alexander Fleming.**

Match the term with its definition.

9. _B_ Kills unwanted insects
10. _D_ Kills unwanted plants
11. _H_ Kills unwanted fungus
12. _C_ Farming without man-made chemicals
13. _A_ Growing plants without soil
14. _E_ Medicine to kill bacteria
15. _F_ Medicine to encourage natural defenses
16. _G_ Some of these plants have natural medicinal value

CHALLENGE QUESTIONS

Short answer:

17. What are two conditions that inhibit enzyme reaction rate? **Heat and decreased pH (increased acidity).**
18. What condition most promotes decomposition? **Darkness and/or warmth.**
19. What are two controversies surrounding organic farming? **Are organic foods healthier? Can organic farms produce as much as non-organic farms? Can organic farms really control pests? Are GMOs bad for us? Is organic farming better for the environment?**
20. Briefly explain how chemotherapy works to treat cancer. **The chemicals target reproducing cells and prevent them from completing reproduction, thus killing the cells. Cancer cells are quickly reproducing so are killed faster than other cells.**

UNIT 7

APPLICATIONS OF CHEMISTRY

LESSON 29 — PERFUMES

WHAT'S THAT SMELL?

SUPPLY LIST

Jar with a lid   Rubbing alcohol   15 whole cloves

Supplies for Challenge: (several of the following) Ginger root   Mint leaves   Peppermint oil
Cinnamon sticks   Dried fruit   Flower petals   Allspice   Almond extract   Vanilla extract

WHAT DID WE LEARN?

- What is a perfume? **A liquid with a pleasing smell.**
- What must be removed from flower petals to make perfume? **The fragrant oil.**
- Describe the two main methods for removing oil from flower petals. **With solvent extraction, a solvent is used to dissolve the oils, then the solvent is allowed to evaporate. With steam distillation, steam is used to vaporize the oil, then both the oil and water condense and the oil is skimmed off the top of the water.**

TAKING IT FURTHER

- Why should you test a new perfume on your skin before you buy it? **The scent of the perfume in the bottle may not be the same as it is on your skin. The alcohol in the bottle may mask the true scent. So put some on your skin and see how it smells once the alcohol has evaporated.**
- Why wasn't it necessary to use one of the methods described in the lesson to make your homemade perfume? **As the cloves soaked in the alcohol, the scent particles slowly moved into the alcohol from the cloves. This is a very slow process. The methods described in the lesson greatly speed up the process for commercial production of perfume.**

LESSON 30 — RUBBER

DO YOU HAVE A RUBBER BAND?

SUPPLY LIST

Balloon   Rubber band   Permanent marker

WHAT DID WE LEARN?

- What is natural rubber made from? **Latex from a rubber tree.**
- What is synthetic rubber made from? **Petroleum—oil.**
- What is vulcanization? **The process of adding sulfur to rubber to make it elastic in all types of weather.**
- What is a polymer? **A long chain of molecules connected together.**

TAKING IT FURTHER

- Why is it difficult to recycle automobile tires? **The vulcanization process makes the rubber very long lasting, but it also makes it hard to break down the molecules so recycling is difficult.**
- What advantages and disadvantages are there to using synthetic rubber instead of natural rubber? **Synthetic rubber is cheaper than natural rubber; however, it requires petroleum, much of which America must import from other countries.**

LESSON 31 PLASTICS

THE WONDER MATERIAL

SUPPLY LIST

Copy of "Chemical Word Search"

Supplies for Challenge: Borax   White glue   Cornstarch   Markers

CHEMICAL WORD SEARCH

```
S A L T V U K A T M A T O M E
C S T R I S Y N T H E T I C I
C P E R F U M E O R E A D F U
A E S S A K T V I X Y J I D Y
R P H O T O S Y N T H E S I S
B L R P S B O N D I U W T G Z
O A G O W T H E I S C N I E M
H S U L T B N M C D F H L S I
Y T V Y N E W B A S E B L T Z
D I S M Q U I I T A T T A I S
R C R E J E P N O B L I T O M
A R K R E B B U R C A L I N Y
T V U L C A N I Z A T I O N S
E A A C I D D H G Y T R N Y O
W P R U F R E S O P M O C E D
```

WHAT DID WE LEARN?

- What is plastic? **A substance made from polymers that are derived from petroleum.**
- What was celluloid, the first artificial polymer, made from? **From cellulose that comes from cotton plants.**

- What is the difference between thermoplastic and thermosetting resin? **Thermoplastics will become soft when reheated, thermosetting resin plastic will not.**

TAKING IT FURTHER

- Name three ways that plastic is used in sports. **Plastic or vinyl balls, artificial rubber soles on running shoes, plastic hooks to hold soccer nets in place, polyester sports clothes, and many other uses.**

- What advantages do plastic items have over natural materials? **Many plastic items are stronger, more flexible, and longer lasting than their natural counterparts.**

LESSON
32

FIREWORKS

IS IT THE FOURTH OF JULY?

SUPPLY LIST

Construction paper or tagboard   Various colors of glitter   Glue
Supplies for Challenge: Epsom salts (found in the medicine section of the store)
Potassium chloride (used as a salt substitute-may be found in the spice section of grocery store)
Several pinecones   Borax and calcium chloride (may be found with laundry/cleaning supplies)
Copper sulfate (found where swimming pool supplies are sold)   Table salt
Several containers in which to soak the pinecones

WHAT DID WE LEARN?

- What are the key ingredients in a fireworks shell? **The chemical that releases the light, black powder for the explosion, and fuses to light the powder.**

- Why does a fireworks shell have two different black powder charges? **One charge lifts the shell into the air and the other charge blasts the shell open.**

- How do fireworks generate flashes of light? **When the blasting charge explodes, the energy released forces electrons in the chemicals into higher energy levels. When the electrons return to their normal energy levels, they release energy in the form of light.**

- What determines the color of the firework? **The chemical compound that is packed inside.**

TAKING IT FURTHER

- How can a firework explode with one color and then change to a different color? **Two different chemicals are packed in the shell and ignited at different times.**

- Why would employees at a fireworks plant have to wear only cotton clothing? **Nylon, polyester, silk, and other fabrics can build up a static charge. This could be very dangerous when working around black powder because a static discharge could ignite the powder.**

LESSON 33 ROCKET FUEL

DO YOU NEED A ROCKET SCIENTIST?

SUPPLY LIST

Balloon   String   Soda straw   Tape

WHAT DID WE LEARN?

- What is combustion? **A chemical reaction that produces great amounts of heat.**
- What two elements are combined in most modern rocket fuel? **Oxygen and hydrogen.**
- What compound is produced in this reaction? **Water/steam.**
- How does combining oxygen and hydrogen produce lift? **The reaction takes place at very high temperatures—heating the atoms to very high temperatures and thus very high speeds. These molecules exit the engine at great speeds, thus producing lift because of Newton's third law of motion.**
- What is Newton's Third Law of Motion? **For every action there is an equal and opposite reaction.**

TAKING IT FURTHER

- Why is oxygen and hydrogen a better choice for rocket fuel than kerosene was? **The end product of the reaction of oxygen and hydrogen is steam, and the end product of kerosene combustion is carbon dioxide. Water is lighter than carbon dioxide so it can move faster. The faster the molecules are moving when they leave the rocket engine, the more lift they produce.**

QUIZ 7 APPLICATIONS OF CHEMISTRY

LESSONS 29–33

Briefly explain how chemistry is used in the making of each of the following items.

1. Perfume: **Solvent extraction or steam distillation is used to extract the scent molecules from flowers. These are then combined with alcohol to form perfume.**
2. Rubber: **Sulfur is added to rubber/latex and then the mixture is heated to form molecules that are strong and flexible.**
3. Plastic: **Long flexible polymers are formed from petroleum, and then heated and molded into plastic.**
4. Fireworks: **Energy is added to chemical compounds to excite the electrons through explosions. When electrons return to their normal levels, they release light. Chemistry is also used in the combustion reaction of the black powder.**
5. Rocket fuel: **Liquid hydrogen and oxygen are combined at high temperature to produce the combustion reaction that provides the needed thrust for lifting a rocket.**

Mark each statement as either True or False.

6. _**T**_ Vulcanization makes rubber useful in most temperatures.
7. _**F**_ Rubber is made from cellulose.

8. _F_ A polymer is a very short molecule.

9. _T_ Today, synthetic rubber is more widely used than natural rubber.

10. _F_ Perfume smells the same in the bottle as on your skin.

11. _T_ Latex is a natural polymer.

12. _T_ Bakelite was the first useful plastic.

13. _T_ Plastic is an important product in American life.

14. _T_ Fireworks are different colors because of different chemical compounds used.

15. _F_ Recipes for fireworks are freely shared.

16. _F_ Kerosene and carbon dioxide are common rocket fuels today.

17. _T_ Newton's Third law of motion is important in rocket design.

18. _T_ Combustion is a chemical reaction that produces large amounts of heat.

CHALLENGE QUESTIONS

Mark each statement as either True or False.

19. _F_ Scents smell the same on every person.

20. _T_ Silk is a natural polymer.

21. _T_ A milk protein can be used as a glue.

22. _F_ Creating polymers is very difficult.

23. _F_ Flames are always the same color.

24. _T_ Sodium chloride burns with a yellow flame.

25. _T_ Hypergolic rocket fuel is not very common.

26. _T_ Solid rocket engines must use up all of their fuel once they are ignited.

27. _T_ Liquid rocket fuel is used in most space rockets.

28. _F_ It is harder to control the rate at which cryogenic fuel burns than the rate at which solid rocket fuel burns.

29. _T_ Borax, glue, and cornstarch can form a polymer.

30. _F_ Lac is a polymer produced by silkworms.

LESSON 34

FUN WITH CHEMISTRY

UNDERSTANDING CHEMICAL REACTIONS

FINAL PROJECT SUPPLY LIST

Milk (not skim)   Paper towels   Food coloring   Water soluble markers   Liquid dish soap
White glue   Disposable baby diaper   Liquid starch   Scissors   Plastic zipper bags
Eyedropper   Copy of "Fun With Chemistry" worksheet

FUN WITH CHEMISTRY WORKSHEET

•   **See the explanation for each activity in the student manual, page 135.**

WHAT DID WE LEARN?

- What was your favorite chemical reaction? **Answers will vary.**
- Why did you like that reaction? **Answers will vary.**

TAKING IT FURTHER

- What do you think will happen if you use skim milk in the first activity? **There are very few fat molecules in skim milk, so adding the soap will make little difference. The colors will eventually mix, but at a much slower rate.**
- What colors would you expect to see separate out of orange ink? Brown ink? **Orange is a combination of yellow and red. Brown is a combination of yellow, red, and blue.**
- Why is it important not to inhale the sodium polyacrylate from the diaper? **Evan a small amount of this chemical will absorb a lot of water, so it can irritate your lungs and your eyes by drying them out.**

ATOMS & MOLECULES

LESSONS 1–34

For each pair of elements, write I if they are most likely to form an ionic bond, C for covalent bond, or M for metallic bond.

1. _I_ Na + Cl
2. _C_ H_2 + O
3. _C_ O + O
4. _I_ K + Br
5. _M_ Al + Al
6. _I_ Mg + O
7. _C_ C + O_2
8. _M_ Ag + Ag
9. _M_ Cu + Cu

Note: elements on opposite sides of the periodic table are likely to form ionic bonds; elements that are both metals (from the left side) will form metallic bonds; elements from the right side (non-metals) usually form covalent bonds.

Fill in the blanks with the terms from below.

10. A _catalyst_ can be used to speed up a chemical reaction.

11. The products of an _exothermic_ reaction have a higher temperature than the reactants.

12. The products of an _endothermic_ reaction have a lower temperature than the reactants.

13. An _enzyme_ is a catalyst that increases the rate of digestion.

14. An acid and a base combine to form a _salt_.

15. A substance is a/an _acid_ if it releases H^+ ions when dissolved in water.

16. A substance is a/an _base_ if it releases OH^- ions when dissolved in water.

- Draw and label a model of a helium atom, which has an atomic number of 2 and an atomic mass of 4.

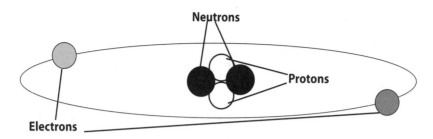

- Choose one of the following topics and briefly explain how chemistry plays a role in it: Farming; Medicine; The nitrogen cycle **Farming: Nitrogen and other chemicals are used up in the growing of crops so chemical fertilizers or other methods must be used to replace them. Also, insecticides, herbicides, and fungicides are all chemicals used to improve crop yield. Medicine: Chemicals are used to change the chemical reactions in the body to improve health. Nitrogen cycle: Nitrogen is used by plants, passed on to animals for their use, and then returned to the soil by decomposers.**

Match the term with its definition.

17. _B_ Natural rubber is made from

18. _D_ Synthetic rubber is made from

19. _E_ A long flexible chain of molecules

20. _A_ Process that makes rubber strong and flexible

21. _F_ A natural polymer found in plants

22. _C_ Process of burning that releases large amounts of heat

Short answer:

23. List three characteristics of a metal. **Silvery, solid, malleable, ductile, conduct electricity.**

24. List three characteristics of a non-metal. **Not shiny, poor conductor, usually gas, brittle if solid.**

25. Explain the chemical reaction involved in your favorite experiment from this book. **Answers will vary.**

CHALLENGE QUESTIONS

26. Use the periodic table of the elements to complete the following chart.

Element	Symbol	Atomic #	Atomic mass	# electrons	# protons	# neutrons
Iron	Fe	26	55.85	26	26	30
Potassium	K	19	39.1	19	19	20
Mercury	Hg	80	200.5	80	80	120 or 121
Krypton	Kr	36	83.8	36	36	48

Fill in the blanks with the words from below.

27. Temperature can increase the _**reaction rate**_ of a chemical reaction.

28. The elements in the first column of the periodic table are _**alkali metals**_.

29. The elements in the last column of the periodic table are _**noble gases**_.

30. The elements in the center of the periodic table are _**transition metals**_.

31. The elements in the second column of the periodic table are _**alkali-earth metals**_.

32. A bioceramic that does not react with the body is a/an _**inert ceramic**_.

33. A bioceramic that dissolves in the body is a/an _**resorbable ceramic**_.

34. An acid is a _**proton donor**_.

35. A base is a **_proton acceptor_**.

36. Molecules that have water bonded to them are **_hydrates_**.

37. A **_homogeneous catalyst_** is in the same phase as the reactants.

38. A **_heterogeneous catalyst_** is in a different phase from the reactants.

LESSON 35

CONCLUSION

APPRECIATING OUR ORDERLY UNIVERSE

SUPPLY LIST

Bible

RESOURCE GUIDE

Many of the following titles are available from Answers in Genesis (www.AnswersBookstore.com).

Suggested Books

Structure of Matter by Mark Galan in the *Understanding Science and Nature* series from Time-Life Books—Lots of real-life applications of chemistry

Inventions and Inventors series from Grolier Educational—Many interesting articles

Molecules by Janice VanCleave—Fun activities

Chemistry for Every Kid by Janice VanCleave—More fun activities

Science Lab in a Supermarket by Bob Friedhoffer—Fun kitchen chemistry

Science and the Bible by Donald B. DeYoung—Great biblical applications of scientific ideas

200 Gooey, Slippery, Slimy, Weird & Fun Experiments by Janice VanCleave—More fun activities

Elements of Faith by Richard Duncan— meaningful insights and spiritual applications from the periodic table of the elements

Suggested Videos

Newton's Workshop by Moody Institute—Excellent Christian science series; several titles to choose from

Field Trip Ideas

- Visit the Creation Museum in Petersburg, Kentucky
- Visit a greenhouse or hydroponics operation to see the use of chemicals with plants
- Tour a battery store to learn about different types of batteries
- Visit a film processing plant to learn about chemicals in film processing or photo printing
- Visit a pharmacy
- Tour an injection molding plant to learn more about plastics
- Visit a farm to learn about the use of chemicals in farming

CREATION SCIENCE RESOURCES

Exploring the World Around You by Gary Parker—More detailed look at different aspects of ecology

Answers Book for Kids Four volumes by Ken Ham with Cindy Malott—Answers children's frequently asked questions

Creation: Facts of Life by Gary Parker—Good explanation of the evidence for creation

The Young Earth by John D. Morris PhD—Evidence for a young earth

The New Answers Books 1 & 2 by Ken Ham and others—Answers frequently asked questions

Zoo Guide and *Aquarium Guide* by Answers in Genesis—A biblical look at animals, including extinction, defense/attack structures, biomes, and stewardship

MASTER SUPPLY LIST

The following table lists all the supplies used for *God's Design for Chemistry & Ecology: Properties of Atoms & Molecules* activities. You will need to look up the individual lessons in the student book to obtain the specific details for the individual activities (such as quantity, color, etc.). The letter *c* denotes that the lesson number refers to the challenge activity. Common supplies such as colored pencils, construction paper, markers, scissors, tape, etc., are not listed.

Supplies needed (see lessons for details)	Lesson	Supplies needed (see lessons for details)	Lesson
Alka-Seltzer	17c, 20c	Glitter	32
Ammonia (clear)	23, 23c	Gloves (leather and cotton)	10
Antacid tablets or liquid	23, 24	Grass and other plants	27
Baking soda	1, 12c, 17, 23	Hydrogen peroxide	19
Balloons (latex)	30, 33	Jar (with lid)	17, 20, 22c, 29
Banana	26c	Lemon juice	19, 22, 24
Battery (9-volt)	12c	Marbles	25c
Bible	35	Margarine	8, 28
Borax	31c, 32c	Marshmallows (mini, colored)	11, 12, 13
Bread	28	Matches	9, 10, 17
Cabbage (red/purple)	21	Mentos candies	1c
Candle	9, 10, 17	Milk (not skim)	22, 34
Copper sulfate (available at swimming pool supply store)	32c	Modeling clay	15c, 17
Cornstarch	31c	Oil (olive)	12c
Cups (clear)	17c	Oil (vegetable)	8
Cups (foam)	20c	Paper clips	22c
Cups (paper)	12c	Paper towels	34
Diaper (disposable)	34	Peanut butter	8
Dish soap	10c, 23, 34	Pennies	22c
Dry ice	10	Pineapple juice (fresh, not frozen)	25c
Eggs	7, 20	Pinecones	32c
Epsom salt	15, 32c	Plant food	27
Eyedropper	34	Plaster of Paris	15c
Flashlight with battery	6	Plastic zipper bags	26c, 34
Food coloring	34	Plate (ceramic)	9
Garlic powder	28	Polymer clay (Femo, Sculpey, etc.)	16
Gelatin	25c	Potassium salt (in spice section)	32c
Geode (optional)	15	Potato	19
Ginger ale	28	Rubber band	30
		Rubbing alcohol	29

Supplies needed (see lessons for details)	Lesson
Salt	12c, 15, 22c, 24, 32c
Silver object (tarnished)	14
Silver polish/tarnish remover	14
Soft drink (lemon lime)	22
Soft drink (diet 2-liter bottle)	1c
Spices (ginger root, mint leaves, cinnamon sticks, allsp ice, cloves, peppermint oil, almond extract, etc.)	29, 29c
Starch (liquid)	34
Steel wool without soap	10c, 20
Stopwatch	17c, 20c
Straw	33
String	33

Supplies needed (see lessons for details)	Lesson
Sugar	12c
Swabs	24
Tape (electrical or duct)	6, 33
Test tubes	10c
Thermometer	20
Toothpaste (with fluoride)	7, 23
Toothpicks	1c, 11, 12, 13
Vinegar	1, 7, 17, 20, 22, 23c, 25c
Water (distilled)	12c, 23c
Wire (copper)	6, 12c
Yeast	26c

WORKS CITED

"Alexander Fleming." http://www.pbs.org/wgbh/aso/databank/entries/bmflem.html.

"Bioceramics." http://www.azom.com/details.asp?ArticleID=1743.

Brice, Raphaelle. *From Oil to Plastic*. New York: Young Discovery Library, 1985.

"Buckyballs." http://scifun.chem.wisc.edu/chemweek/buckball/buckball.html.

"Charles Goodyear and the Strange Story of Rubber." *Reader's Digest*. Pleasantville, N.Y.: January 1958.

"Charles Martin Hall." http://www.geocities.com/bio-electrochemistry/hall.htm.

"Charles Martin Hall and the Electrolytic Process for Refining Aluminum." http://www.oberlin.edu/chem/history/cmharticle.html.

"Chemotherapy, What it is, How it Helps." http://www.cancer.org/docroot/ETO/content/ETO_1_2X_Chemotherapy_What_It_Is_How_It_Helps.asp.

Chisholm, Jane, and Mary Johnson. *Introduction to Chemistry*. London: Usborne Publishing, 1983.

Cobb, Vicki. *Chemically Active Experiments You Can Do at Home*. New York: J.B. Lippincott, 1985.

Cooper, Christopher. *Matter*. New York: Dorling Kindersley, 1992.

"Development of the Periodic Table." http://mooni.fccj.org/~ethall/period/period.htm.

"Diapers, the Inside Story." http://portal.acs.org/portal/fileFetch/C/CSTA_014946/pdf/CSTA_014946.pdf.

Dineen, Jacqueline. *Plastics*. Hillside: Enslow Publishers Inc., 1988.

Dunsheath, Percy. *Giants of Electricity*. New York: Thomas Y. Crowell Co., 1967.

"Enzyme Chemistry." http://www.math.unl.edu/%7Ejump/Center1/Labs/EnzymeChemistry.pdf?id=11897.

"Farming , Food and Biotechnology." *Inventions and Inventors*. 2000.

Galan, Mark. *Structure of Matter - Understanding Science and Nature*. Alexandria: Time-Life Books, 1992.

Helmenstine, Anne Marie, Ph.D. "Chemistry." http://chemistry.about.com.

"Historical Development of the Periodic Table." http://members.tripod.com/~EppE/historyp.htm.

"How and Why Science in the Water." *World Book*. 1998.

"How Does a Halogen Light Bulb Work?" http://home.howstuffworks.com/question151.htm.

Hughey, Pat. *Scavengers and Decomposers: The Cleanup Crew*. New York: Atheneum, 1984.

Julicher, Kathleen. *Experiences in Chemistry*. Baytown: Castle Heights Press, 1997.

Kuklin, Susan. *Fireworks: the Science, the Art, and the Magic*. New York: Hyperion Books for Children, 1996.

"Medicine and Health." *Inventions and Inventors*. 2000.

Morris, John D., Ph.D. *The Young Earth*. Green Forest: Master Books, 1998.

Newmark, Ann. *Chemistry*. New York: Dorling Kindersley, 1993.

Parker, Gary. *Creation Facts of Life*. Colorado Springs: Master Books, 1994.

Parker, Steve. *Look at Your Body - Digestion*. Brookfield: Copper Beech books, 1996.

"Penny For Your Thoughts." http://www.tryscience.org/experiments/experiments_pennythoughts_athome.html.

Pinkerton, J.C. "Alexander Fleming and the Discovery of Penicillin." http://nh.essortment.com/alexander-flemin_rmkm.htm.

"Polymers: They're Everywhere." http://www.nationalgeographic.com/resources/ngo/education/plastics/nature.html.

Richards, Jon. *Chemicals and Reactions*. Brookfield: Copper Beech books, 2000.

Saari, Peggy and Stephen Allison, Eds. *Scientists: The Lives and Works of 150 Scientists*. U.X.L An Imprint of Gale, 1996.

"Silly Putty." http://www.chem.umn.edu/outreach/Sillyputty.html.

Solids, Liquids, and Gases. Ontario Science Center. Toronto: Kids Can Press, 1998.

Student Activities in Basic Science for Christian Schools. Greenville: Bob Jones University Press, 1994.

Thomas, Peggy. *Medicines from Nature*. New York: Twenty-First Century Books, 1997.

VanCleave, Janice. *Chemistry for Every Kid*. New York: John Wiley and Sons, Inc., 1989.

VanCleave, Janice. *Molecules*. New York: John Wiley and Sons, Inc., 1993.

"Vulcanized Rubber." http://inventors.about.com/library/inventors/blrubber.htm.

"WebElements Periodic Table of the Elements." http://www.webelements.com/index.html.

Wile, Jay. *Exploring Creation with Chemistry*. Anderson: Apologia Educational Ministries, 2003.